Health 129
健康的马儿

Healthy Horses

Gunter Pauli

冈特·鲍利 著
凯瑟琳娜·巴赫 绘
章里西 译

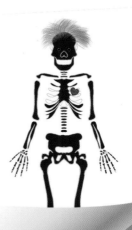

学林出版社
www.xuelinpress.com

丛书编委会

主　任：贾　峰

副主任：何家振　闫世东　郑立明

委　员：牛玲娟　李原原　李曙东　李鹏辉　吴建民
　　　　彭　勇　冯　缨　靳增江

特别感谢以下热心人士对译稿润色工作的支持：

王必斗　王明远　王云斋　徐小帖　梅益凤　田荣义
乔　旭　张跃跃　王　征　厉　云　戴　虹　王　逊
李　璐　张兆旭　叶大伟　于　辉　李　雪　刘彦鑫
刘晋邑　乌　佳　潘　旭　白永喆　朱　廷　刘庭秀
朱　溪　魏辅文　唐亚飞　张海鹏　刘　在　张敬尧
邱俊松　程　超　孙鑫晶　朱　青　赵　锋　胡　玮
丁　蓓　张朝鑫　史　苗　陈来秀　冯　朴　何　明
郭昌奉　王　强　杨永玉　余　刚　姚志彬　兰　兵
廖　莹　张先斌

目录

健康的马儿	4
你知道吗？	22
想一想	26
自己动手！	27
学科知识	28
情感智慧	29
艺术	29
思维拓展	30
动手能力	30
故事灵感来自	31

Contents

Healthy Horses	4
Did you know?	22
Think about it	26
Do it yourself!	27
Academic Knowledge	28
Emotional Intelligence	29
The Arts	29
Systems: Making the Connections	30
Capacity to Implement	30
This fable is inspired by	31

野马们成群结队地驰骋在蒙古的大草原上。春天来了，冰雪中冒出了青青绿草，花儿也开始绽放。初来乍到的马驹们吸引了一只山地绵羊的注意。

"在春天迎来年轻一代，感觉棒极了。你们的孩子生得真俊，祝贺你们！"山地绵羊说道。

Wild horses are running across the steppe of Mongolia. Spring has come! Fresh grass is sprouting through the snow, and the first flowers blossom. A mountain sheep is looking at the new foals that have just arrived.

"It is wonderful to see the next generation arrive in the spring. Congratulations on the beautiful babies," the sheep says.

初来乍到的马驹们吸引了一只山地绵羊的注意。

A mountain sheep is looking at the new foals.

春天还是来了!

Spring is here at last!

"谢谢夸奖。我很庆幸我们挺过了又一个寒冬。在那么冷的天里风餐露宿真是太遭罪了,世上所有的马都不该受这样的折磨!"母马答道。

"然而春天还是来了,风也变得和煦起来了。"

"但我更喜欢秋天的风,吹着更健康。"

"Thank you so much. I am glad that we made it through another winter. No horse in the world should weather such freezing winds without any shelter," the mare responds.

"Spring is here at last, and the winds are warmer."

"Well, I prefer the healthy autumn winds."

"不过那会儿的风吹着不冷吗?"山地绵羊问道。

"但是,夏天里高温把土地加热了不少,所以秋风吹起来只会给我们带来凉爽和舒适,也让我们为熬过漫长的冬天做些准备。你说对吗?"

"你的意思是,春天的风吹着不健康咯?"

"But aren't the winds cold then?" Sheep asks.
"But as the summer heat has warmed up the land, the winds in the fall are soothing, bringing us comfort, preparing us for the long winter, you see."
"Are you claiming that the spring wind is not healthy?"

秋风吹起来凉爽又舒适

The winds in the fall are soothing

把寒冷吹到我们骨子里

Keeps the cold in our bones

"冬天漫长又寒冷,空气暖和起来几周后,土地也还是冻着的。所以春天的风很凉,还可能让我们生病,能把寒冷吹到我们骨子里,而且没法帮我清空嗓子里的痰。"

"但至少你们的骨骼比任何其他动物的骨骼都强健。"

"The winters are long and cold, and the land is still frozen for weeks after the air has warmed up. That is why a cold spring breeze could make us sick. It keeps the cold in our bones and does not help to get the mucous out of my throat."

"At least you have very strong bones, stronger than any other animal."

"我们的骨骼是由大量钙质组成的。"

"所有生物的骨骼都是由钙质组成的啊。"

"可不管是寒冬还是盛夏,我们一年到头都在奔跑。我们享有的这片天空能给我们充足的阳光,让我们的骨骼无比结实。"

"Our bones are made out of packs of calcium."
"All bones are made of calcium."
"But we run all year round, both during the winter freeze and the summer sun. Our open skies are rich in healthy sun rays that make our bones so solid."

……骨骼是由大量钙质组成的。

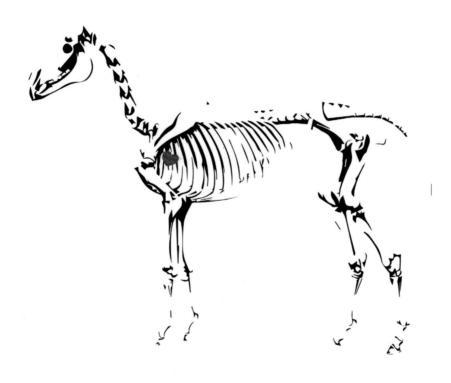

... bones are made out of packs of calcium.

花样繁多的素食食谱!

The most diverse vegetarian diet!

"所以你们才能当野马，能这么奔跑啊！"山地绵羊笑道。

"你说得好像很简单似的，可我们还得吃得杂才行。你知不知道，我们吃的植物有400种！我们还爱舔食岩石上的苔藓，或者偶尔来点儿蘑菇。"

"你们的骨骼这么强健，一定是有着花样繁多的素食食谱呀。"

"That is why you can run … and be a wild horse!" Sheep giggles.
"You make it sound so simple, but we do need a rich diet. Do you know that we eat four hundred different kinds of plant? We also love to lick lichen from the rocks and enjoy a mushroom once in a while?"
"You must have the most diverse vegetarian diet to make such strongest bones."

"要是人类也知道如何让骨骼强壮就好了!他们只知道吃肉,却忽略了只有强壮的骨骼才能撑起强壮的身体。"

"是呀,我听说人类的骨骼强度一年不如一年了。有些人不跌跟头,都能发生骨折。"

"是吧,他们最好学着点我们的饮食习惯,或许还可以试着吃一些我们的骨头。"

"If people only knew how they could get strong bones too! They only think about eating meat – and forget that a strong body depends on a strong skeleton."

"True, I hear that people's bones are getting weaker every year. Some break bones even without falling."

"Well, they had better learn from our diet, and perhaps start eating some of our bones."

强壮的骨骼撑起强壮的身体

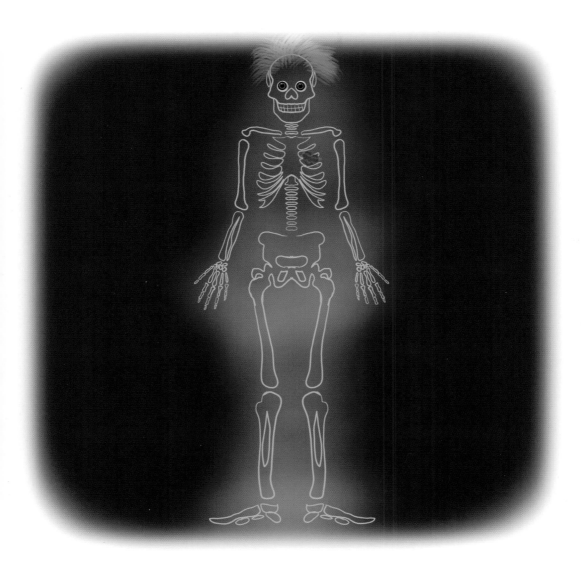

A strong body depends on a strong skeleton

花几个小时炖煮马骨,做一锅老火靓汤……

Cook bones for hours and make a soup ...

"狗才吃骨头,人不吃的。"

"我不是说人类应该啃马的骨头。我是说人们应该花几个小时炖煮马骨,做一锅老火靓汤。一匹马身上的硬骨大约有三十公斤,一般都直接扔掉或者烧成灰了,挺浪费的。"

"毕竟人类降生在地球上还是最近的事儿嘛,他们还处在学习的过程中。"

"People do not eat bones, dogs eat bones."

"I am not saying that people should gnaw on horse bones. I mean that they should cook them for hours and make a great soup out of them. A horse has about thirty kilos of solid bones and these are usually just thrown away, or even burnt to ashes."

"Well, humans are recent arrivals on earth. They are still learning."

"希望他们学快点儿吧,多吃蔬菜、多喝骨头汤。否则不用多久他们就会被自己的体重压垮啦。"

"我们可不希望发生那种事,对吧?"

……这仅仅是开始!……

"Let's hope they learn fast: to eat more veggies and have more bone soup. Or they may soon collapse under their own weight."

"And we don't want that to happen, now do we?"

... AND IT HAS ONLY JUST BEGUN!...

……这仅仅是开始!……

...AND IT HAS ONLY JUST BEGUN!...

你知道吗？

The Mongolian Horse, also known as the Przewalski horse, is a rare and endangered wild horse native to the steppes of Central Asia. It is the only true wild horse left on earth.

蒙古野马也叫普氏野马，原产于中亚地区的大草原，是一种珍稀而濒危的野马。蒙古野马是世界上仅存的真正意义上的野马。

During the winter, Przewalski horses dig for plants that grow beneath ice and snow. For water, they eat snow. Przewalskis are accompanied by the domesticated Mongol horses, which remained largely unchanged since the time of Genghis Khan.

在冬天普氏野马会挖掘埋藏在冰雪之下的植物吃，并通过吃雪来摄入水分。普氏野马与已经驯化的蒙古马为伴，不过这些驯化的蒙古马自从成吉思汗时代后就几乎没有什么变化了。

Mongolian soldiers preferred to ride lactating mares because they could use their milk. In times of desperation, soldiers would slit a minor vein in the horse's neck and drain some blood into a cup to prevent dehydration.

蒙古士兵更爱骑哺乳期的母马，因为可以顺便饮用马奶。到了弹尽粮绝时，士兵们会切开马脖子上的小静脉血管，放血到杯子中来饮用以抵御脱水。

There are five kinds of herd animals recognised in Mongolia: horses, camels, yaks, sheep, and goats. The horse is the most valued. There are 500 Mongolian words to describe the traits of a horse.

蒙古官方认可的畜牧畜类有5种：马、骆驼、牦牛、山地绵羊、山羊。马是其中最受重视的，蒙古语中描述马的特征的词语多达500个。

The mare births a foal in 20-45 minutes, faster than any mammal. The foal is usually up and nursing within two hours of birth. Horses sleep standing up as excessive ground contact causes skin, muscle, and bone trauma.

母马分娩一匹小马驹只需要 20 到 45 分钟，比任何其他哺乳动物都快。小马驹一般出生后 2 小时内就可以站立、喝奶了。马站着睡觉，因为过多与地面接触会导致皮肤、肌肉和骨骼损伤。

Competition with livestock, hunting, capture for zoological collections, military activities, and harsh winters are the main causes of the decline in the Przewalski horse population. By the 1950s, only 12 individual Przewalski horses were left in the world.

目前普氏野马的族群数量明显减少，主要原因包括与家畜的生存竞争、捕猎、为动物学研究进行的捕捉、战争以及冬季的严寒。20 世纪 50 年代，全世界仅剩 12 匹普氏野马。

In 1977, the Foundation for the Preservation and Protection of the Przewalski horse was founded in Rotterdam (Netherlands) by Jan and Inge Bouman. The reintroduced horses reproduced, and its status was changed from "extinct in the wild" to "endangered".

1977年,扬·褒曼和英格·褒曼夫妇在荷兰鹿特丹成立了普氏野马保护基金会。人们诱导仅存的普氏野马交配,已经使该物种的濒危等级从"野外灭绝"变为"濒危"。

Sunlight is an excellent natural source of Vitamin D3, which stimulates the absorption of calcium and magnesium through the skin and is essential for maintaining strong and healthy bones. Sunlight triggers an increase in the feel-good brain chemical serotonin.

阳光是获取维生素 D_3 的绝佳天然渠道,有助于人们通过皮肤吸收钙、镁以保持骨骼强健。阳光也能提升脑部化学物质血清素的含量,使人产生愉悦感。

Think About It / 想一想

If walking and exercise outside in the sun is the best for our health, why do we sit inside and work behind a computer for such long periods?

如果在室外阳光下走路、锻炼对健康是最有好处的,那我们为什么要在室内坐着,在电脑前工作那么长时间?

Do you think it is possible for us to reverse the extinction of certain animals when we only find a few species left in zoos?

我们现在只能在动物园里找到为数不多的动物品种。据此你认为我们有可能逆转动物的灭绝趋势吗?

Which animal would be your preferred symbol of strength, and why?

你认为哪种动物更适合作为力量的象征?为什么?

Would you eat bones, provided you did not have to chew on them?

如果骨头不用嚼就能碎,你会吃它吗?

Are people worried about their bone strength? Ask around and find out if those who are worried about this know that calcium alone cannot do the job. Calcium pills may put more calcium in your veins, causing a stroke instead of thickening your bones. This exercise is to build up arguments that simply eating more calcium is not enough. Good health also requires enough exercise and improved eating habits, while spending more time outdoors. Turns this into a logical argument, with some jokes added. One of the greatest ways to convince people of something is to make them laugh, and they will even laugh at themselves.

人们会担心骨骼不够强健吗？问问周围的人，看看他们知不知道，单纯补钙是没法让骨骼强壮的。过量服用钙片可能会导致中风。想要身体健康，人还需要进行足够的室内外运动并改善饮食习惯。请把这些观点转换成有逻辑的论点，再加上一些笑话。让人发笑是说服他人的最佳手段之一。

TEACHER AND PARENT GUIDE

学科知识
Academic Knowledge

生物学	普氏野马有66条染色体，其他马类只有64条染色体；种马负责聚拢马群、驱赶敌人和保卫的工作，而母马领导整个族群；蒙古马的奶可加工成马奶酒，是蒙古的国民饮料。
化学	骨骼会在身体呈碱性时强化，在身体呈酸性时变脆；人出生时大部分骨骼是软骨，之后硬化成骨，这个过程叫作骨化；维生素D是一组脂溶性类固醇，在人体内负责钙、铁、镁、磷酸盐和锌的肠道吸收。
物理	马用耳朵表达兴趣，从它们耳朵尖的朝向可以观测到它们的注意力焦点；马不能通过嘴巴呼吸，它们长长的鼻孔和鼻道能够过滤、加温、加湿空气；马的气管较长，需要利用重力清除气管中的黏液，因此它们吃草时头朝下；马的心电传导系统能够将心率从静止提高到每分钟300次左右；紫外线照射皮肤可以合成维生素D_3。
工程学	通过食品工程可以将维生素添加到食品中并增添新营养素。
经济学	国民的骨骼健康会影响整个经济运行——医生、健康产业、健美和体育产业、管理膳食的营养学家、生物教师、复健理疗师以及急救人员都需要关注骨骼健康；维生素D_3来源于羊毛脂。
伦理学	20世纪30年代，英国医学研究委员会的政策是不对科学发现申请专利保护，因为该机构相信医学研究的结果应该免费向所有人提供。
历史	1928年，德国哥廷根大学教授阿道夫·温道斯被授予诺贝尔化学奖，以表彰他对固醇的构成以及固醇和维生素关系研究的贡献。
地理	蒙古马是一种蒙古高原的古老马种，历史上横扫欧亚的蒙古士兵就是骑着这种马。
数学	光在真空中的速度约为300 000km/s，当光进入空气、玻璃等介质中时，传播速度会变慢。
生活方式	骨质疏松症是一种疾病；不健康的生活习惯如在室内久坐，持续避免阳光照射皮肤，运动量小，蔬菜、坚果和水果摄入不足，等等，会导致骨骼变得非常脆弱；吸烟会提高人体酸度，阻碍骨骼愈合；缺乏维生素D会导致缺钙，并引发佝偻病、肥胖症等。
社会学	马是蒙古士兵的交通工具，还是他们的食物及饮料来源，并能用来制作盔甲、鞋、装饰品、弓弦、绳索等；他们还依靠马来取火、运动，并且作为生活中音乐、狩猎等娱乐活动乃至精神力量的来源；他们相信自己死后，马会成为他们冥间的坐骑。
心理学	骨骼的弱化会导致试探性、警惕性行为增多，这些行为会削弱信心，使运动的平衡性减弱，增加摔倒及骨折的风险；骨质疏松症导致脊柱出现弯曲，阻碍直立站立，可能加重抑郁；马术治疗。
系统论	太阳是维持我们健康的关键，但我们却因为畏惧皮肤癌而远离太阳。

教师与家长指南

情感智慧
Emotional Intelligence

绵羊　　绵羊对马表达了共情,她在祝贺母马诞下小马时充满了感情。她转移到更为轻松的天气话题,营造了友好的氛围。当不明白母马的逻辑时,她敢于追问更多细节。她承认马的优点:强健的骨骼。当母马分享了一些显而易见的事情时,她隐晦地坚持让母马给出进一步的解释。她很放松并且会打趣。她对周围事物的观察细致认真。绵羊反应很快(她指出人们不吃骨头),而且宅心仁厚,对于不知该做些什么也不懂得如何生活的人类表示了宽容。

母野马　　母马懂得感恩,和绵羊交谈的内容不只是例行公事的客套话,而是分享了母子共同经历的苦难。她和绵羊关于天气的闲聊变成了一堂关于健康和气候的公开课。她花时间解释为什么对马而言秋天的风比春天的风更好。当话题转移到骨骼强度时,母马指出了太阳的重要性以及阳光有益健康的特性。她提醒绵羊不要把事物简单化。她认为人类不懂得如何关爱自己的骨骼,并建议人类向马学习,并且她对马骨的浪费感到惋惜。母马以一种戏谑的口吻结束了对话。

艺术
The Arts

让我们一起看一部关于马的电影。可选的影片很多,比如《黑骏马》《沙漠骑兵》《梦想奔驰》《马语者》《自由飞奔》和《弗莉卡》。如果你想听歌颂马的音乐,考虑考虑"美国"乐队的《无名马》或者是派蒂·史密斯的《马群》;如果你喜爱古典乐,那么柴可夫斯基钢琴组曲《四季》中的第11首《十一月——在马车上》是不错的选择,罗西尼《〈威廉·退尔〉序曲》第四乐章关于骑兵进军的描写也很精彩。

TEACHER AND PARENT GUIDE

思维拓展
Systems: Making the Connections

常年在野外生存的野马,为习惯了当代生活方式的我们上了重要的一课,揭示了这种生活方式的消极影响。

骨骼的生长贯穿生命始终,健康饮食可以让身体维持碱性,而饮用碳酸饮料、摄入糖分、吸入酸性的空气甚至吸烟,都会导致我们的骨骼逐渐溶解。这是因为为了代偿低pH值,机体会将具有高pH值的钙释放进血液。临床研究可能不支持这一观点,但整体医学要求我们正视人体的过高酸负荷。海洋的酸化会对珊瑚产生破坏,而软体动物在合成贝壳过程中会被迫消耗更多能量。同样的机理也适用于人体:为了补偿机体受到的损害,确保身体机能维持原有水平,我们需要消耗比之前多得多的能量。

我们应该像马一样,经常在田地里漫步,吃健康的蔬菜,在大自然中奔跑、运动,定期沐浴阳光。不仅是在炎炎夏日,我们在冬天和秋天也应该晒太阳,因为紫外线能让我们的皮肤合成维生素D。

但这样的建议让人们产生了更多的疑惑。为避免患皮肤癌的风险,许多人使用防晒霜,在窗户上装紫外线过滤器,让自己免受光照。然而现实是,皮肤癌的患病率没有下降,骨骼疾病愈发普遍了。简单的自我保护是远远不够的。除了保护自己免受不利因素的损害,我们还应当对我们有益的事物采取更加接纳的态度。

动手能力
Capacity to Implement

你听说过马术治疗吗?这种疗法针对有特殊需求的儿童和成人,用来促进他们的身心健康。马对患有自闭症、痴呆症、智力发育迟缓、唐氏综合征、抑郁症和脑损伤的患者能产生积极影响,这些患者能与马和谐相处并最终熟练掌握各种马术技巧。找一处对有特殊需求儿童开放的马棚,做一个简单的练习:试着不用手接触就让某匹马走出围栏。你会发现鼓掌、喊叫和吹口哨往往不奏效。所以,我们应该用正确的方式激励特殊儿童,强迫他们或者对他们大喊大叫、拍手都是不合适的。

教师与家长指南

故事灵感来自

英格·褒曼
Inge Bouwman

　　1972 年，在与丈夫扬·褒曼度蜜月期间，英格·褒曼在布拉格动物园见到一匹被关在拥挤围栏里的普氏野马，夫妻俩开始想象马群获得自由、重返蒙古高原的场景。他们了解到这种马全世界只剩下几百匹，于是决定投入所有的时间、精力和金钱来拯救这个物种。他们为所有尚存的马创建了一本种系谱，基于遗传学规律成功培育了一支优良的种系。1992 年，在世界自然基金会和荷兰政府的支持下，第一批普氏野马抵达了蒙古哈斯台国家公园和戈壁沙漠。丈夫去世后，英格没有停止她的事业。在 72 岁时，她决定将这份事业传给在蒙古中部长大的生物学家乌斯呼吉日嘎拉·多吉。受这个传奇故事启发，中国在西北的卡拉麦里自然保护区建立了自己的育种中心。

图书在版编目（CIP）数据

健康的马儿：汉英对照 /（比）冈特·鲍利著；（哥伦）凯瑟琳娜·巴赫绘；章里西译．—上海：学林出版社，2017.10
（冈特生态童书．第四辑）
ISBN 978-7-5486-1253-7

Ⅰ．①健… Ⅱ．①冈… ②凯… ③章… Ⅲ．①生态环境－环境保护－儿童读物－汉、英 Ⅳ．① X171.1-49

中国版本图书馆 CIP 数据核字（2017）第 143424 号

© 2017 Gunter Pauli
著作权合同登记号　图字 09-2017-532 号

冈特生态童书
健康的马儿

作　　者——	冈特·鲍利
译　　者——	章里西
策　　划——	匡志强　张　蓉
特约编辑——	隋淑光
责任编辑——	汤丹磊
装帧设计——	魏　来
出　版——	上海世纪出版股份有限公司 学林出版社
	地　址：上海钦州南路81号　电话/传真：021-64515005
	网　址：www.xuelinpress.com
发　　行——	上海世纪出版股份有限公司发行中心
	（上海福建中路193号　网址：www.ewen.co）
印　　刷——	上海丽佳制版印刷有限公司
开　　本——	710×1020　1/16
印　　张——	2
字　　数——	5万
版　　次——	2017年10月第1版
	2017年10月第1次印刷
书　　号——	ISBN 978-7-5486-1253-7/G·479
定　　价——	10.00元

（如发生印刷、装订质量问题，读者可向工厂调换）